MASTER
MAGIC PUZZLES

T.N. MAHESH

Publisher:
INDIRA PUBLISHERS
F2, Sree Lakshmi Apartments,
24/67, Valmiki Street, East Tambaram, Chennai - 600 059.
Cell: 97890 73754.
Website: www.magicsquarepuzzles.com
Email: indirapublishers@gmail.com
info@magicsquarepuzzles.com

Master Magic Puzzles©

Edition	:	October 2023
Author	:	T.N. Mahesh, 9840662780.
Publisher	:	Indira Publishers©
Book Size	:	7 in x 10 in
Pages	:	98
Design	:	Bramma Designs, Chennai - 41. Ph: 9841304824
Publisher	:	**INDIRA PUBLISHERS**

INDIRA PUBLISHERS
F2, Sree Lakshmi Apartments,
24/67, Valmiki Street,
East Tambaram, Chennai - 600 059.
Website: www.magicsquarepuzzles.com
Email: indirapublishers@gmail.com
        info@magicsquarepuzzles.com
Mobile: 97890 73754.

How Long Does the Sunlight Take to Reach The Earth?
https://witty-thinker-8124.ck.page/5d580e7bfc

This book is dedicated to
*my Mother **Indira Narasinga Rao,***
whose work fascinated and inspired me to take up
magic squares for study.
But for her abiding interest in magic squares,
I would not have ventured into this field.

You have got to follow the soul;
Wherever it goes, Wherever it goes.

FOREWORD

Dr. R. Gopalan,
Head of the Department of
Chemistry (Retd),
Madras Christian College,
Tambaram,
Chennai – 600 059.

Secretary,
Sri Sankara Vidyalaya,
Pammal.

The book Magic Square Puzzles, authored by T.N. Mahesh, is an interesting and challenging compilation of Fifty Magic Square Puzzles. He calls Magic Square Puzzles as Magic Puzzles.

The Magic Square Puzzles have been formulated by the author based on his research work on this innovative and interesting topic.

His mother Srimathi Indira Narasinga Rao was the motivating factor for this author to pursue this subject with avid interest and dedication.

The author has imparted a modern flavour to this ancient mathematical magic square concept. India has been a mathematical giant in the world since the Vedic times. This hoary tradition is sustained by the contemporary authors such as Mahesh.

The magic number concept is a unique blend of mathematics and religious belief. It was applied in China, India, and also the Arab World for the interpretation of some aspects of religious concepts.

The subtle aspects of mathematics, such as the theory of numbers and arithmetic, are ingrained in the study of magic squares. It is poetry in mathematics.

Solving the puzzles in this book will not only sharpen the skill in mathematics but also provide recreational pleasure for students.

I am confident that this book written with in-depth knowledge and commitment by the author will whet the enthusiasm and creativity of our young mathematicians in schools and colleges. In addition, this book will revive our ancient and celebrated knowledge of mathematics.

I congratulate the author on writing this valuable book and wish him the best in his future endeavour.

R. Gopalan

Register For Free Training For Magic Puzzles
https://forms.gle/sRR1hww9y2gQsvrt9

ENJOY MATHEMATICS

Follow us:  YouTube

Watch The Full Video

https://www.youtube.com/watch?v=hkvRw-6yvi4

1. **What is a Magic Square**
   https://www.youtube.com/watch?v=A-AG19n408g&t=29s

2. **What is a Magic Sum**
   https://www.youtube.com/watch?v=B1UkVJTE2AU&t=7s

3. **Magic Square Formulas**
   https://www.youtube.com/watch?v=Gzq6P6CqyQg&t=9s

4. **How to solve Magic Square Puzzle**
   https://www.youtube.com/watch?v=fmsYYGyQS0g&t=14s

1

2

3

4

PREFACE

I was introduced to the concept of magic squares through the work of my Mother Indira Narasinga Rao, who passed away recently.

My Mother had constructed a 1000x1000 magic square and had found a mention in Limca Book of Records. She was also honored by the Ramanujan Museum, situated in Royapuram, Chennai and was interviewed by a few reputed Tamil TV channels.

The then Director of Ramanujan Museum, Shri. P.K. Srinivasan, asked my Mother to write a book on her discoveries in magic squares and she hesitatingly started the work. I was helping my Mother write the book Magic of Magic Squares and in that process came to know the subject under her guidance.

The magic squares fascinated, intrigued and challenged me to find more about them. But what is a magic square?

Magic square is an array of n2 numbers; each number used only once and arranged in such a way that each row sum, column sum and the diagonal sum of the resultant magic square gives the same total, also called the magic sum.

In short we can say that Magic Square is poetry in the arrangement of numbers.

There are 880 unique magic squares of order 4x4 and my mother worked mainly in the Sree Ramar Chakram or the 646th magic square. She scaled the 4x4 Sree Ramar Chakram to 1000x1000 to find a mention in the Limca Book of Records.

The Sree Ramar Chakram (4x4 magic square) and Seetha Chakram (3x3 magic square) are found in the Hindu almanac and are used to make predictions.

I thought why not scale up all the 880 magic squares of order 4x4 and by God's grace, I was able to do that.

I found a pattern in the emerging magic squares and used those patterns to create a puzzle and I call these Magic Square Puzzles.

My Mother came close to discovering all these concepts but somehow I made it. The genius is hers and I just added a little to her works.

It was my Mother's desire to create a puzzle based on the magic squares and I am happy that I did it.

These puzzles are now being published daily in The Hindu In School, a newspaper catering exclusively to schools, from the reputed daily The Hindu.

I have taken classes on Magic Square Puzzles in several schools and the students simply love this topic. There is a very lively atmosphere in the classes and the teachers say that my classes are not like regular Mathematics classes.

Going by the reaction from the student community, I wonder why it should not be kept as part of the curriculum. If an opportunity comes, I will surely help the authorities to place this in the curriculum.

These puzzles involve both Arithmetic and Logic and the students who try these puzzles will start doing basic arithmetic operations like addition and subtraction in their mind itself over a period of time.

These puzzles also help students overcome their fear of Mathematics because of the playway method.

In fact in the classes that I take for the children, I expressly ask the students to put down their notebook, pencil, etc. and try to solve these puzzles in their mind and the students do that.

NOTE TO PARENTS AND TEACHERS

Even if the children use calculators initially, over a period of time they will start doing these puzzles without the help of calculators, in their mind itself. This process will happen naturally. Most of the students show a marked improvement after solving a few puzzles.

For more of these puzzles visit www.magicsquarepuzzles.com

If you want these puzzles to appear in any Newspaper or Magazine kindly let me know

Wish you a very nice journey through these puzzles.

HOW TO SOLVE THE
MAGIC SQUARE PUZZLE

What is a magic square ?

A magic square is an arrangement of numbers from 1 to n^2 in an nxn matrix with each number occurring exactly once such that the sum of the entries of any row, column and any main diagonal is the same.

For example a 3x3 magic square consists of numbers from 1 to 9, arranged in a 3x3 matrix format, each only once with each row, column and diagonal giving a sum of 15. We call this **magic sum.**

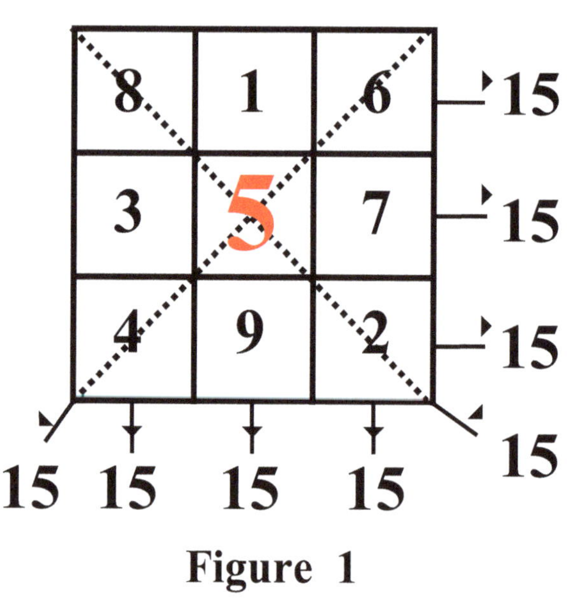

Figure 1

Let us analyze this 3x3 magic square.
The Row1, 8+1+6 = 15,
The Row2, 3+5+7 = 15 and
The Row3, 4+9+2 = 15.

Now columns
Column1, 8+3+4 = 15,
Column2, 1+5+9 = 15 and
Column3, 6+7+2 = 15.

Now Diagonals
Diagonal1, 8+5+2 = 15 and
Diagonal2, 6+5+4 = 15.

All the rows, columns and the diagonals give the same sum of 15, so 15 is called the magic sum. Observe the centre number 5. If you multiply this centre number 5 with the order 3, you get the magic sum 15, (Figure 1).

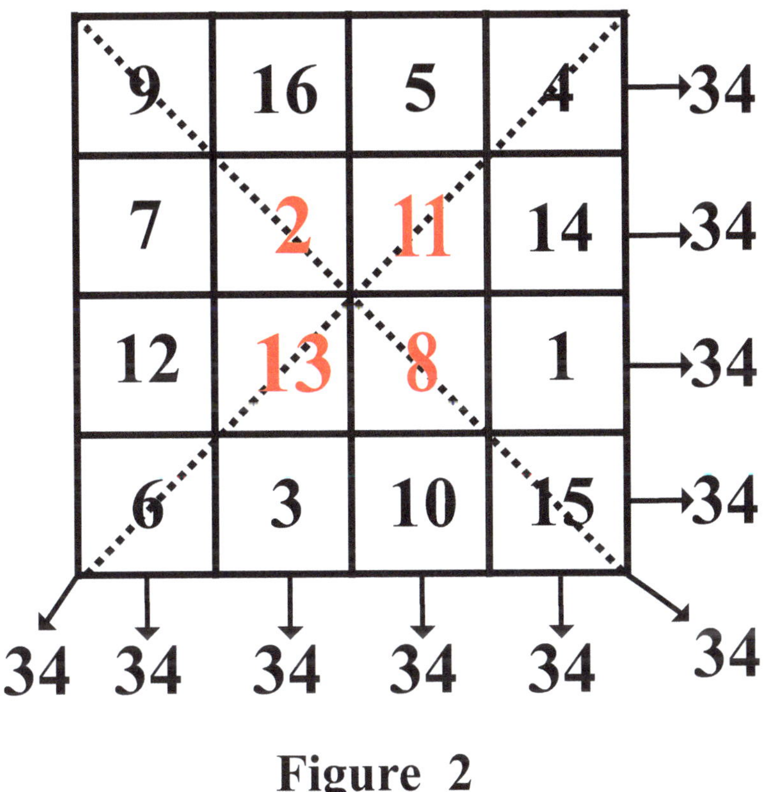

Figure 2

Since each row, column and two diagonals give the sum of 34, we call this as the magic sum. You can add the numbers in each row, column, and diagonal and verify it yourself.

If you add the center 2x2 of the given 4x4 magic square you get the sum of 2+11+13+8 = 34 (Figure 2). This holds good for all the 4x4 magic squares.

In the 3x3 we have only 1 magic square and in 4x4 we have 880 unique 4x4 magic squares.

In the parlance of the magic squares rotation, mirror image etc are not counted as different but the same magic square.

Magic sum for the 3x3 square is 15, 4x4 is 34, 5x5 is 65. For an 8x8 magic square, the magic sum is 260.

A typical 8x8 magic square obtained by using the mimic method (discovered by me) consists of **four 4x4 magic squares.** While the **magic sum of an 8x8 magic square is 260, the magic sum of each 4x4 subsquare marked as A, B, C and D is 130** (Figure 3).

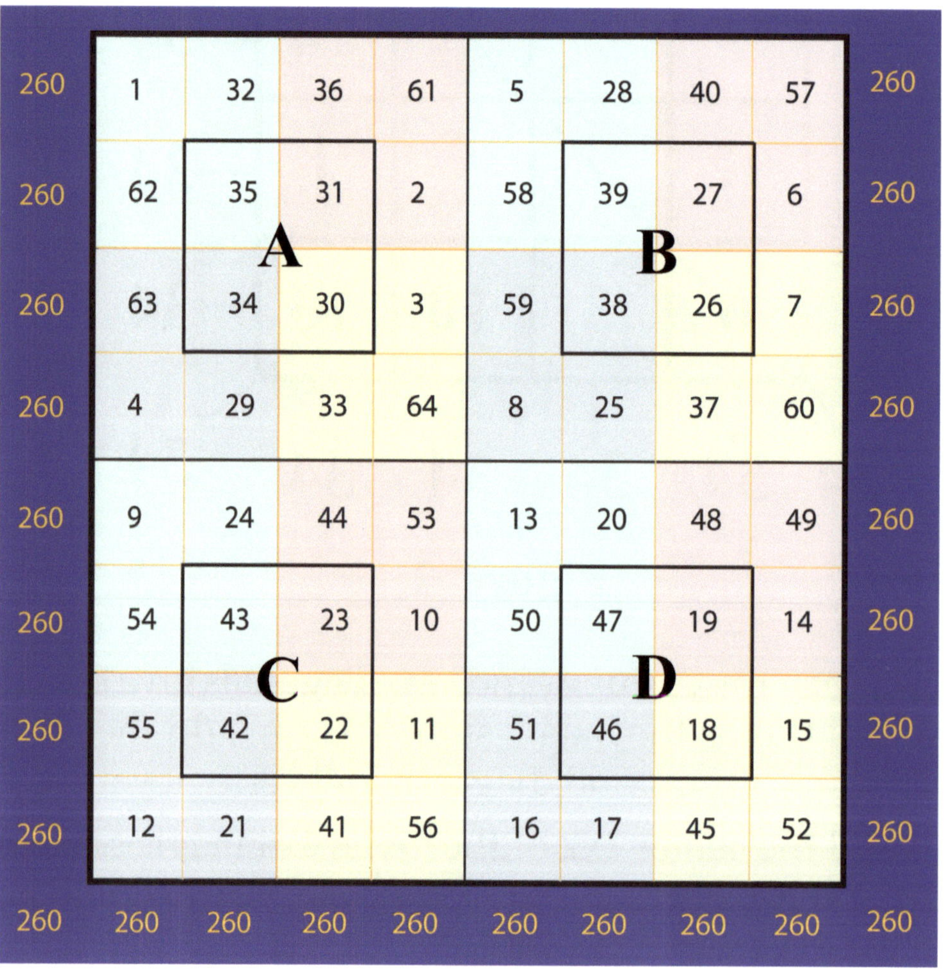

Figure 3

The centre 2x2 of each 4x4 magic square enclosed in a box gives a magic sum of 130.

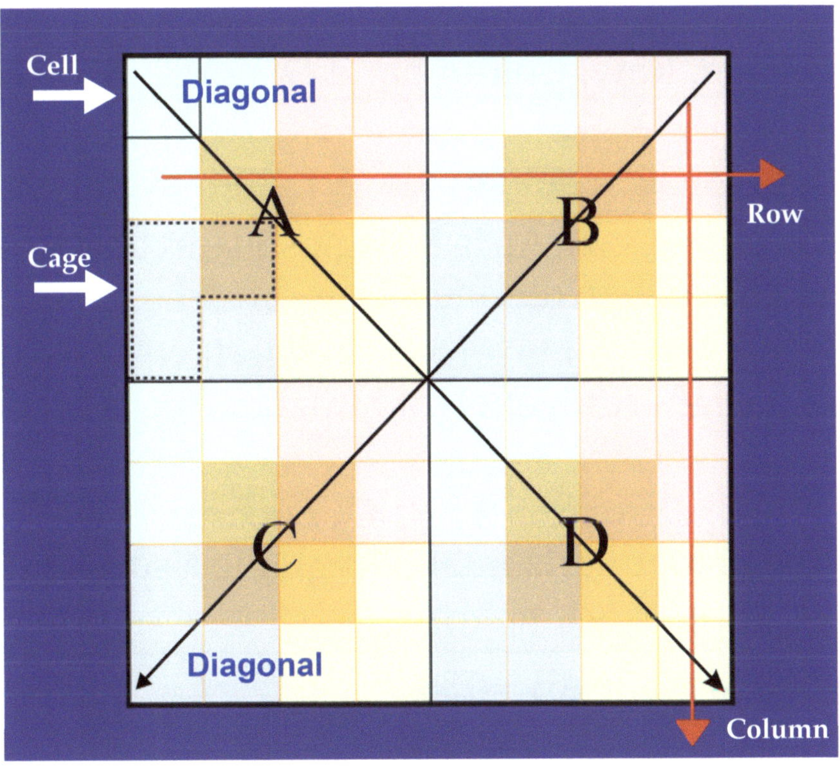

Figure 4

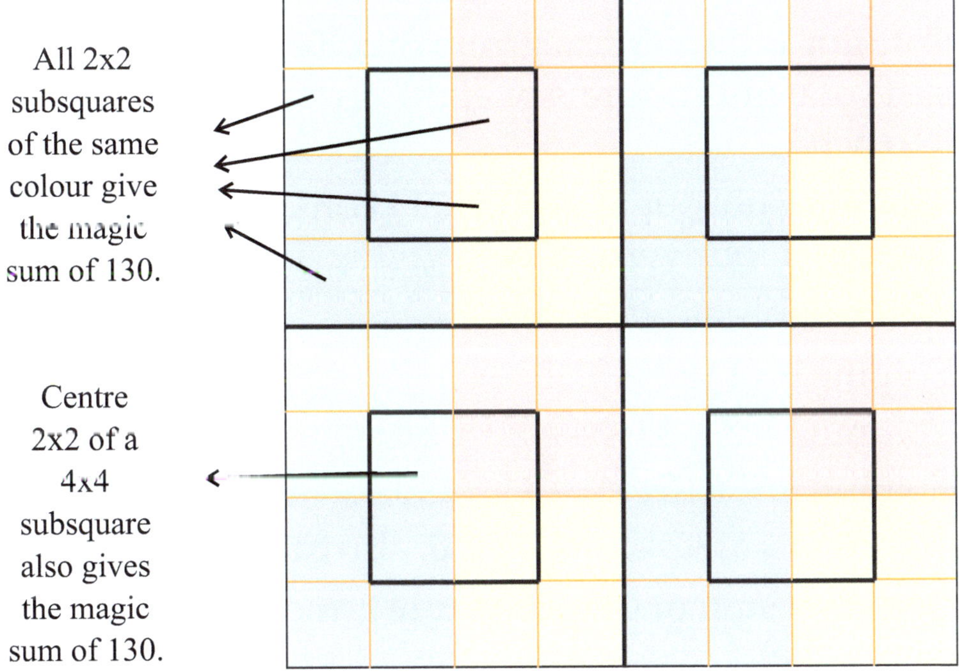

Figure 5

Have a look at the Figure 4, to understand the concepts of Cell, Cage, Row, Column and Diagonal, which we will discuss now.

We will consider **subsquare A** for our examples.

Cell is a single box. An 8x8 magic square consists of 64 cells and the cells contain numbers from 1 to 64.

Row is the horizontal group of cells. If you take the 8x8, it consists of eight rows of eight horizontal cells giving a sum of 260 each.

Any 4x4 subsquare **A, B, C, D** consists of four rows consisting of four horizontal cells giving a sum of 130 each.

For example as per Figure 7,

Row1 = 1+32+36+61 = 130,

Row2 = 62+35+31+2 = 130,

Row3 = 63+34+30+3 = 130,

Row4 = 4+29+33+64 = 130, also called **row rule.**

Column is the vertical group of cells. If you take the 8x8, it consists of eight columns of eight vertical cells giving a sum of 260 each.

Any 4x4 subsquare A, B, C, D consists of four columns consisting of four vertical cells giving a sum of 130 each.

For example as per Figure 7,

Column1 = 1+62+63+4   = 130,

Column2 = 32+35+34+29 = 130,

Column3 = 36+31+30+33 = 130,

Column4 = 61+2+3+64   = 130, also called the **column rule.**

Diagonal sum or diagonal rule is the sum of the diagonal elements like

Diagonal1 = 1+35+30+64 = 130,

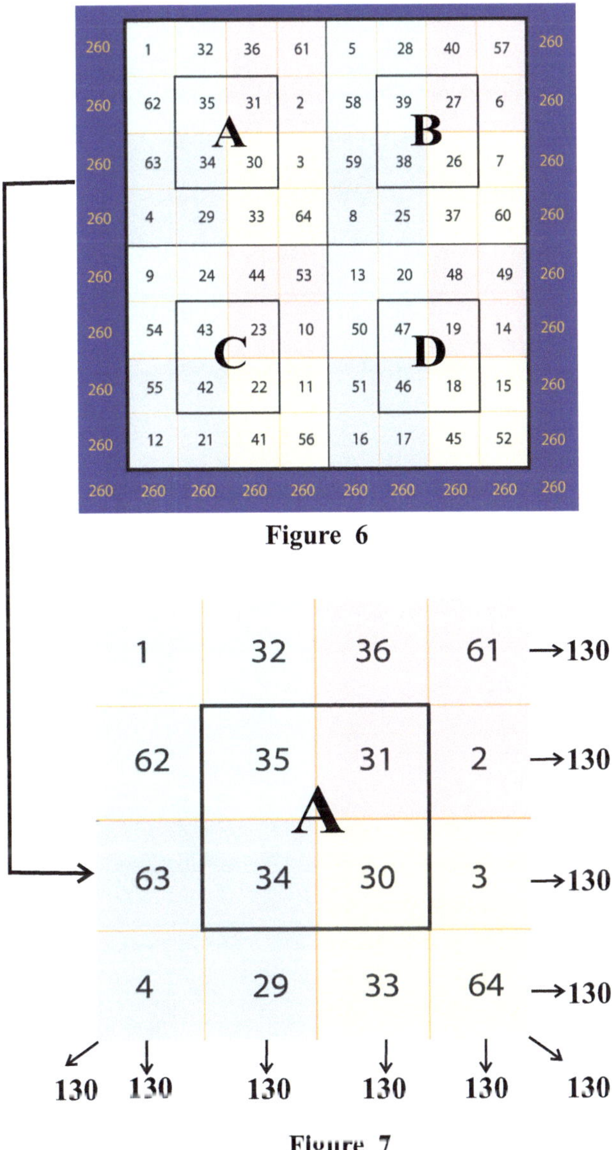

Figure 6

Figure 7

Diagonal2 = 61+31+34+4 = 130, again as per Figure 7.

Cage is the group of cells marked by the perforated or dashed lines.

Cage sum* is the sum of the group of cells within the cage. The cage sum is given in the top left corner of the respective cages. We will see the example for the cage sum later.

Part cage sum is the sum of some of the cells within one or more cages.

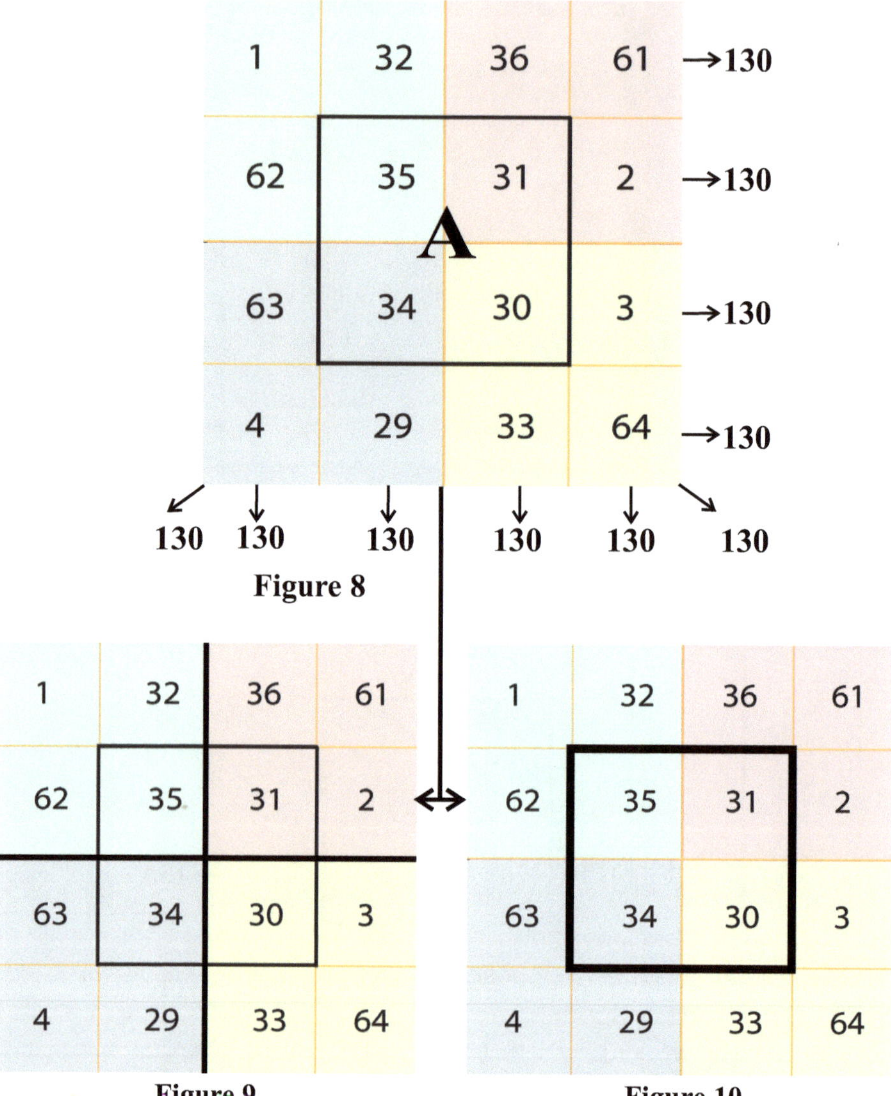

Figure 8

Figure 9

Figure 10

Each 2x2 marked in same color gives the magic sum of 130,

1+32+62+35 = 130,

36+61+31+2 = 130,

63+34+4+ 29 = 130,

30+3+ 33+64 = 130 (Figure 9).

This is called the **color matrix rule.**

The centre 2x2 also gives the magic sum of 130,

35+31+34+30 = 130 (Figure 10). This is called the **box rule.**

These rules hold good for all the 4x4 subsquares marked A, B, C and D. We will now see how the cells are numbered in an 8x8 magic square.

	1	2	3	4	5	6	7	8
A	A1	A2	A3	A4	A5	A6	A7	A8
B	B1	B2	B3	B4	B5	B6	B7	B8
C	C1	C2	C3	C4	C5	C6	C7	C8
D	D1	D2	D3	D4	D5	D6	D7	D8
E	E1	E2	E3	E4	E5	E6	E7	E8
F	F1	F2	F3	F4	F5	F6	F7	F8
G	G1	G2	G3	G4	G5	G6	G7	G8
H	H1	H2	H3	H4	H5	H6	H7	H8

Based upon this numbering we will explain how to solve the magic square puzzle given in Figure 11.

In magic square puzzles it is not necessary you start at a particular cell and proceed in an orderly fashion. You can start anywhere and go anywhere. Sometimes you have to solve some cell in a 4x4 subsquare to find out a number in some other subsquare.

Though the Magic Square Puzzles involve 8x8 magic square, while solving these puzzles we mostly look in the range of 4x4 subsquares.

INSTRUCTIONS

Fill empty cells using numbers from 1 to 64, each only once such that

1) Each row, column and diagonals of each 4x4 subsquare gives the magic sum of 130.

2) Each 2x2 square in the same coloured area gives the magic sum of 130.

3) The centre 2x2 box in each 4x4 subsquare gives a sum of 130.

4) Groups of cells outlined with dotted lines are cages. The cage sum* is given in the top left corner of the respective cages.

Using these instructions, the magic square puzzles can be solved.

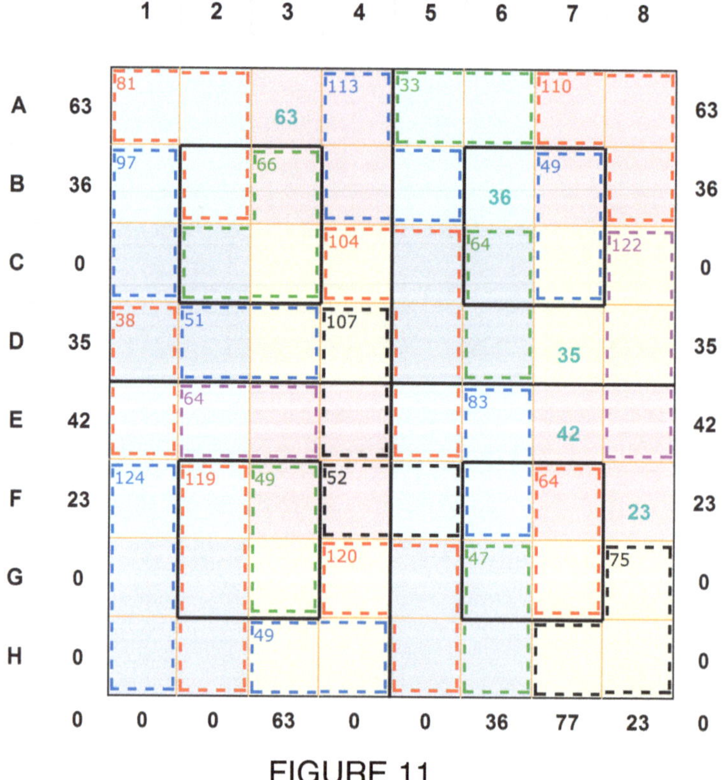

FIGURE 11

We will see few examples for the cage sum.

The cage consisting cells A1+A2+B2 gives the cage sum of 81.

The cage consisting cells A4+B4+B5 gives the cage sum of 113.

The cage consisting cells A5+A6 gives the cage sum of 33.

The cage consisting cells A7+A8+B8 gives the cage sum of 110.

The cage consisting cells B1+C1 gives the cage sum of 97.

The cage consisting cells B3+C2+C3 gives the cage sum of 66.

The cage consisting cells G8+H7+H8 gives the cage
sum of 75, and

The cage consisting cells H3+H4 gives the cage sum of 49.

Likewise there are other cages which go into making this Magic Square Puzzle.

For each puzzle the cages will be different and their cage sums will be different. From these hints you have to find all the 64 numbers in a Magic square Puzzle.

In the above puzzle the numbers 63, 36, 35, 42 and 23 are declared beforehand. Of the 64 numbers in an 8x8 magic square puzzle, these 5 numbers are declared and we have to find the remaining 59 numbers to solve this puzzle.

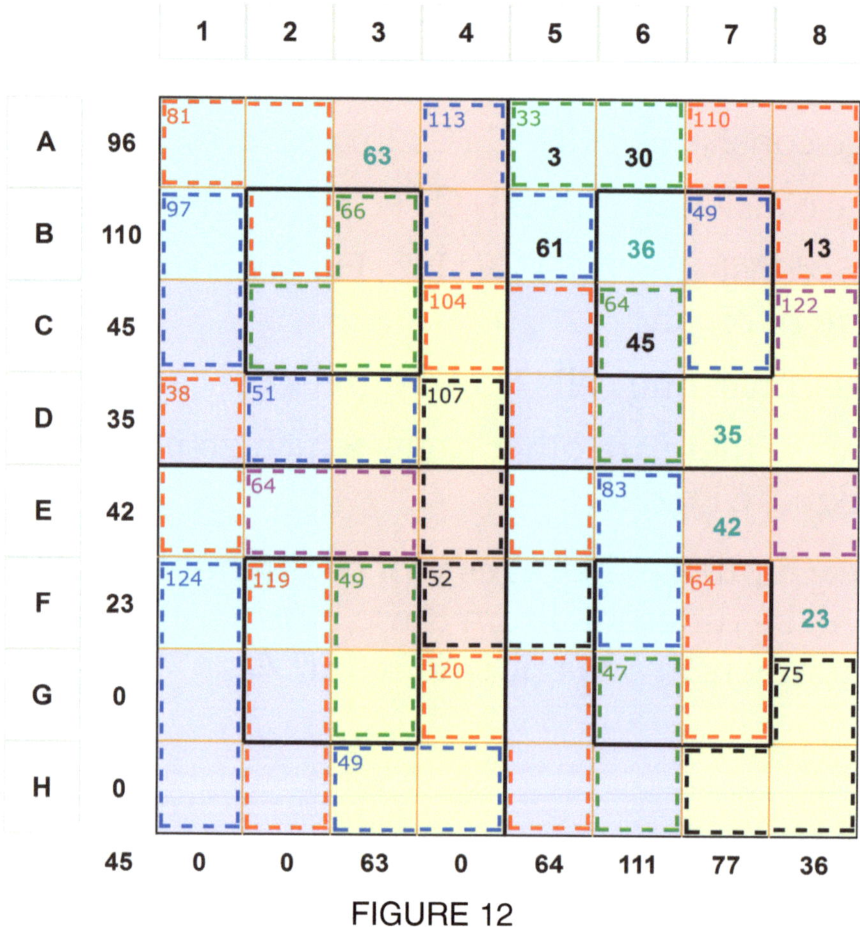

FIGURE 12

The cage sum A5+A6 is 33 and the cage sum A7+A8+B8 is 110. Sum of these two cages (A5+A6) + (A7+A8+B8) = 33+110 = 143. We can write the same as (A5+A6+A7+A8) + B8 = 143. According to **row rule,** A5+A6+A7+A8 = 130. So 130+B8 = 143. Hence B8 = 143-130 = 13 (Figure 12).

What will be A6? We use the **column rule,** A6+B6+C6+D6 = 130. Since B6 = 36 and C6+D6 = 64, A6+36+64 = 130. A6+100 = 130. So, A6 = 130-100 = 30 (Figure 12).

Since the cage sum A5+A6 = 33 and A6 = 30, so, A5 = 33-30 = 3 (Figure 12).

Can we find C6? We have B6 = 36 and B7 + C7 = 49. As per the **box rule,** B6+B7+C6+C7 = 130. So, C6 = 130 - (B6+(B7+C7)). C6 = 130-(36+49) = 130-85 = 45 (Figure 12).

Since we know A5, A6 and B6, it is easy to find out B5, using **color matrix rule.** Since A5+A6+B5+B6 = 130, (A5+A6)+B5+B6 = 130. So 33+B5+36 = 130. So B5 = 130 - (33 + 36) = 130 - 69 = 61 (Figure 12).

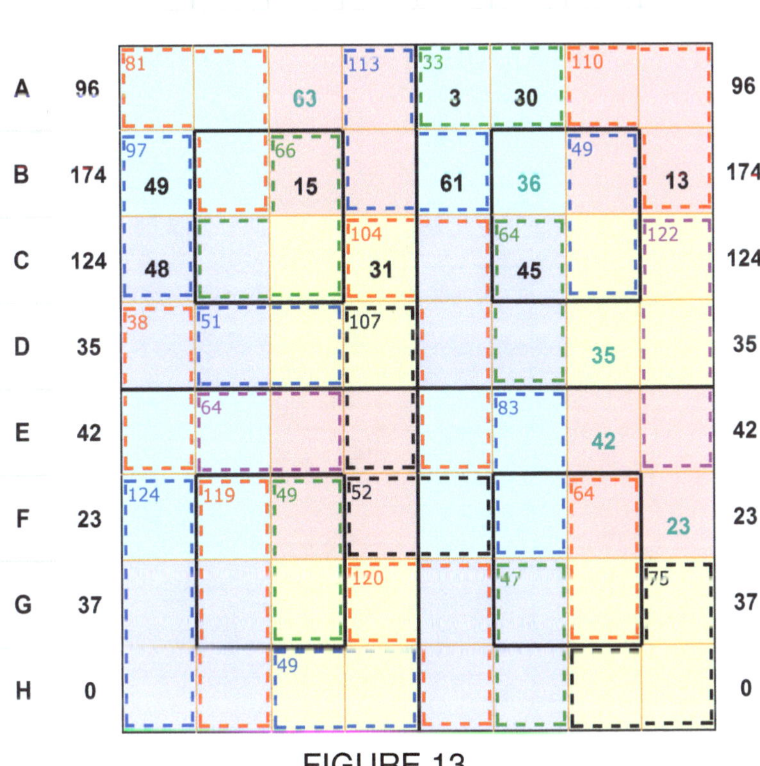

FIGURE 13

We will see how **part sum of a cage** is used in solving a puzzle. In Figure 13, the cage A4 + B4 + B5 = 113. We know B5 = 61. So the **part cage sum** of A4 + B4 = 113 - B5 = 113 - 61 = 52.

We know A3 = 63 and A4+B4 = 52. As per **color matrix rule** A3+A4+B3+B4 = 130. So A3+A4+B4+B3 = 130. 63+52+B3 = 130.

B3+115 = 130. So B3 = 130-115 = 15 (Figure 13).

We can find B1 using the **color matrix rule.** A1+A2+B1+B2 = 130. (A1+A2+B2) = 81, as per the puzzle. So B1 = 130-(A1+A2+B2). So B1 = 130-81 = 49 (Figure 13).

B1+C1 = 97 as per the puzzle. Since B1 = 49, 49+C1 = 97. So C1 = 97-49 = 48 (Figure 13).

What is C4? The cage B3+C2+C3 = 66 and we have B3 = 15. That means, the **part cage sum** of C2 + C3 = 66 - 15 = 51.

Also C1 = 48 and as per the **row rule** C1 + C2 + C3 + C4 = 130. So C1 + (C2+C3) + C4 = 130. 48 + 51 + C4 = 130. That is 99 + C4 = 130. C4 = 130 - 99 = 31 (Figure 13).

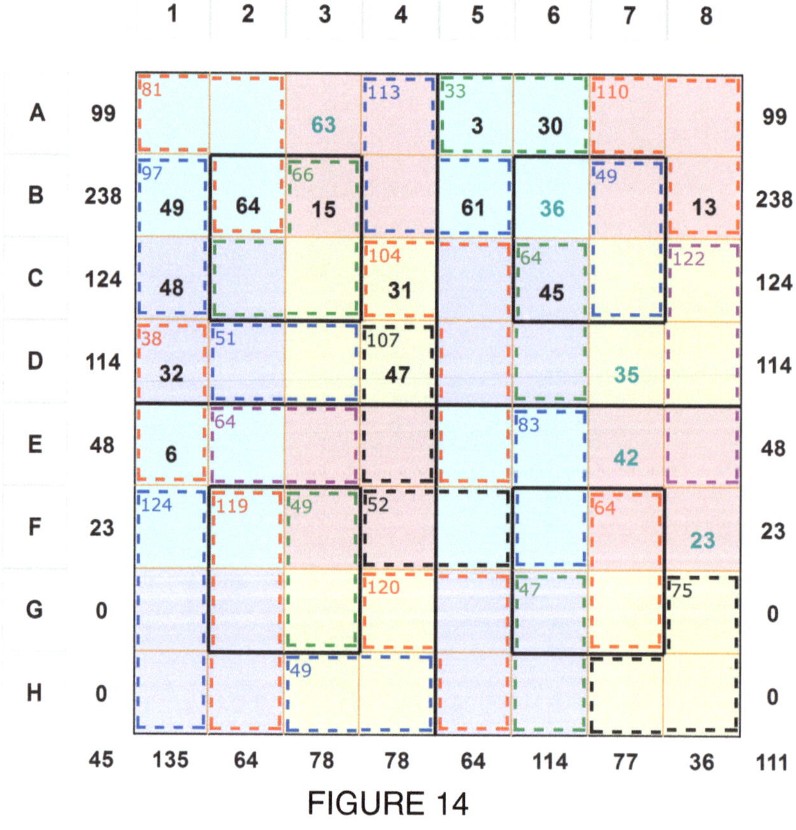

FIGURE 14

As per **box rule** B2+B3+C2+C3 = 130. But the cage B3+C2+C3 = 66. So B2 = 130-(B3+C2+C3). B2 = 130-66 = 64 (Figure 14).

Since the **part cage sum** of A4+B4 = 52 and C4 = 31, we can find D4 applying **column rule,** A4+B4+C4+D4 = 130. 52+31+D4 = 130, so D4 = 130-(52+31). D4 = 130-83 = 47 (Figure 14).

Since F1+G1+H1=124 and applying the **column rule** E1+F1+G1+H1 = 130, we get E1+124 = 130. So E1 = 130-124, so E1 = 6 (Figure 14).

Since cage D1+E1 = 38 and E1 = 6, D1 = 38-E1. D1 = 38-6 = 32 (Figure 14).

We can also find the D1 in the following manner. D4 = 47 and the cage D2+D3 = 51. Applying the **row rule** D1+D2+D3+D4 = 130. D1 = 130-(D2+D3+D4) = 130-(51+47). D1 = 130-98 = 32 (Figure 14).

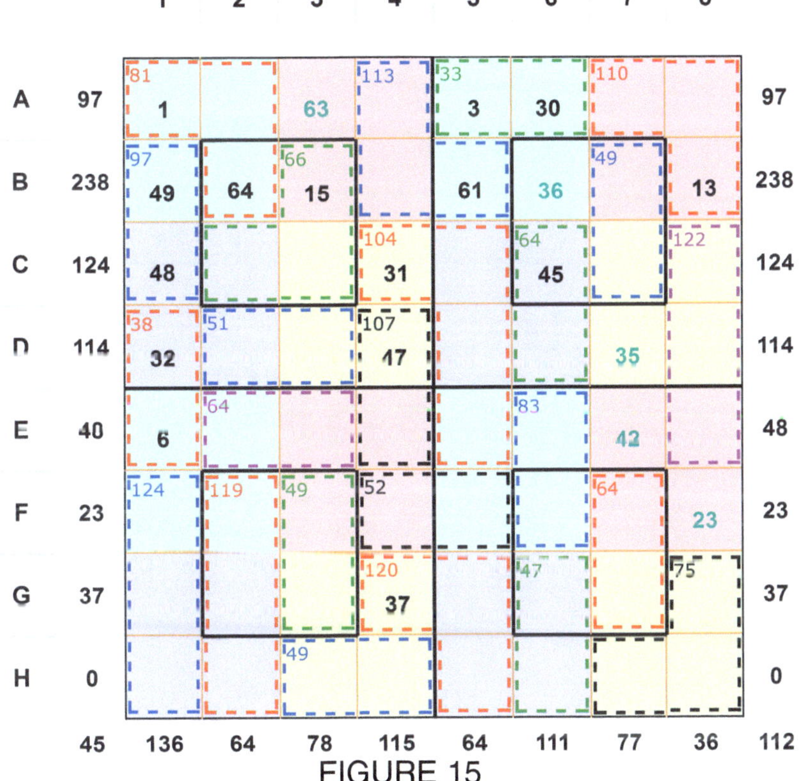

FIGURE 15

We can now find A1, using the **column rule.** Since A1+B1+C1+D1 = 130, A1 = 130-((B1+C1)+D1). B1+C1 = 97 and D1 = 32. So A1 = 130-(97+32). A1 = 130-129 = 1 (Figure 15).

We will now find G4. G5+G6+H5+H6 = 130 as per **color matrix rule.** The cage sum of G4+G5+H5 = 120 and the cage sum of G6+H6 = 47. So G4+ G5+G6+H5+H6 = 120+47 = 167. So G4+130 = 167. G4 = 167-130 = 37 (Figure 15).

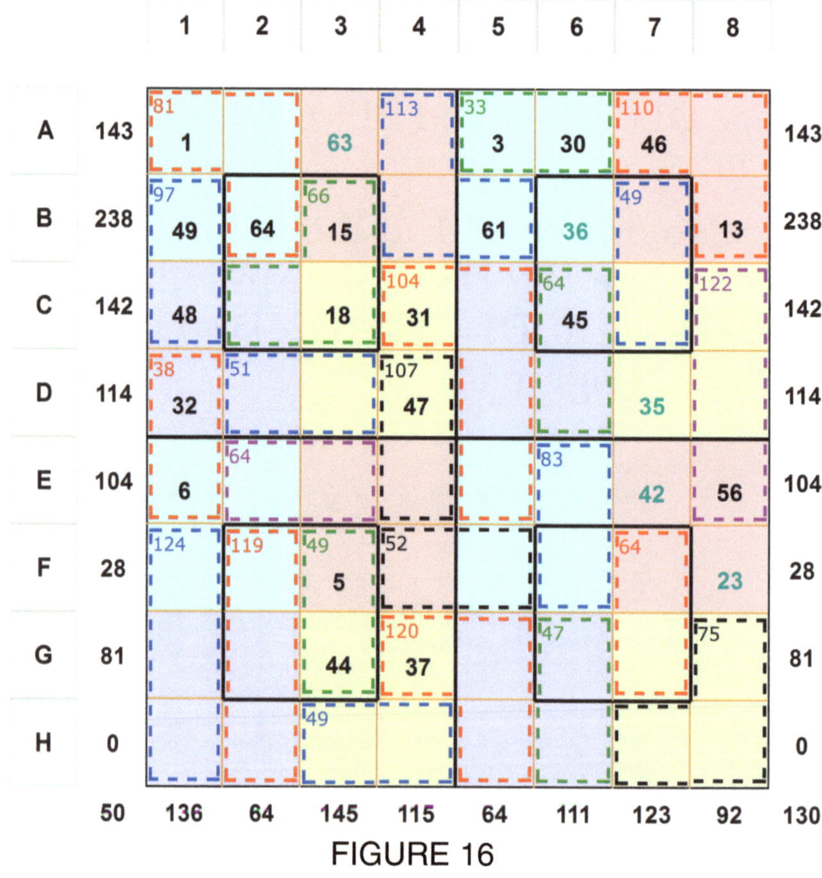

FIGURE 16

We can now find C3 using **diagonal rule.** Since A1+B2+C3+D4 = 130 and we know that A1 = 1, B2 = 64 and D4 = 47, substituting these values we get 1+64+C3+47 = 130. C3+112 = 130. So C3 = 130-112 = 18 (Figure 16).

We can now find G3 and F3. G4=37 and the cage sum of H3+H4 = 49. As per the **color matrix rule** G3+G4+H3+H4 = 130. G3+37+49 = 130. G3+86 = 130. G3=130-86 = 44. Since F3+G3 = 49 and G3 = 44, F3 = 49-G3 = 49-44 = 5 (Figure 16).

We can now find A7. The **cage sum** of B7+C7 = 49 and D7 = 35. Applying the **column rule,** A7+B7+C7+D7 = 130, we get A7+49+35 = 130. A7+84 = 130. A7=130-84 = 46 (Figure 16).

Since A7 = 46 and cage A7+A8+B8 = 110, the **part cage sum** of A8+B8 = 110-A7. So, A8+B8 = 110-46 = 64.

As per the puzzle C8+D8+E8=122. So A8+B8+C8+D8+E8 = 64+122 = 186. As per **column rule** A8+B8+C8+D8 = 130. So, 130+E8 = 186. E8 = 186-130 = 56 (Figure 16).

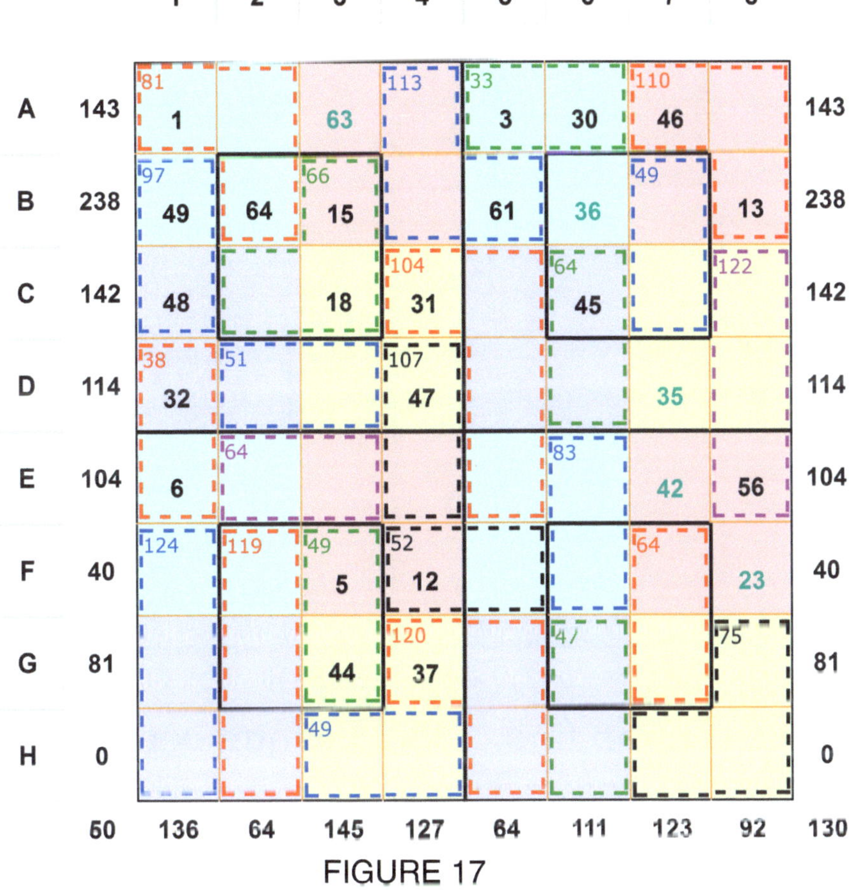

FIGURE 17

We can now find F4 using **part cage sum** method.

The cage sum of F1+G1+H1 = 124 and the cage sum of F2+G2+H2 = 119. The sum of these two cages (F1+G1+H1)+(F2+G2+H2) = 124+119 = 243.

We will find the cage sum of F1+F2. As per the **color matrix rule,** G1+G2+H1+H2 = 130. Since F1+F2+(G1+G2+H1+H2) = 243, so F1+F2 = 243-130 = 113.

Since F3 = 5, using the **row rule** F1+F2+F3+F4 = 130 we get 113+5+F4 = 130. So 118+F4 = 130. F4 = 130-118 = 12 (Figure 17).

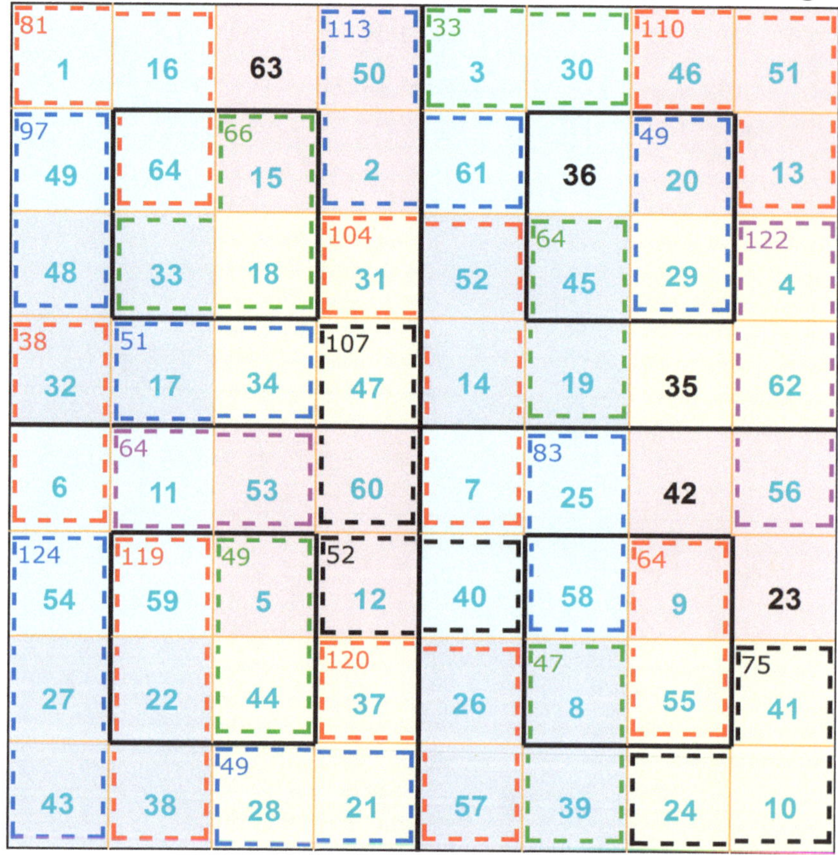

FIGURE 18

Likewise you can find out all the numbers and solve the puzzle.

As you keep doing it, you will feel that you can do simple arithmetic operations like addition and subtraction in the mind itself. You will also be comfortable with basic algebra. All the best on embarking on an interesting and absorbing voyage.

The completed puzzle is given in Figure 18.

TIPS FOR MENTAL CALCULATION

Suppose we have to find 130 - 82, divide the problem into two parts. What is 100-82? It is 18. Then add 30 to 18 to get 48.

To find the difference between 23 and 85, count in 10's from 23 like 33....43...53.....63....73.....83 and add 2. We get 62. By calculating this way we can easily calculate in our mind itself.

Kindly give your rating and review @ amazon.in
https://amzn.eu/d/bgl7VL1

If you want these puzzles to appear in any
Newspaper or Magazine kindly let me know

Email: indirapubllshers@gmail.com
info@magicsquarepuzzles.com

All the puzzles found in this book can be solved using these instructions and hints. But there are more hints or patterns to be found in the magic squares as given in the figures that follow. You can use these also to solve them.

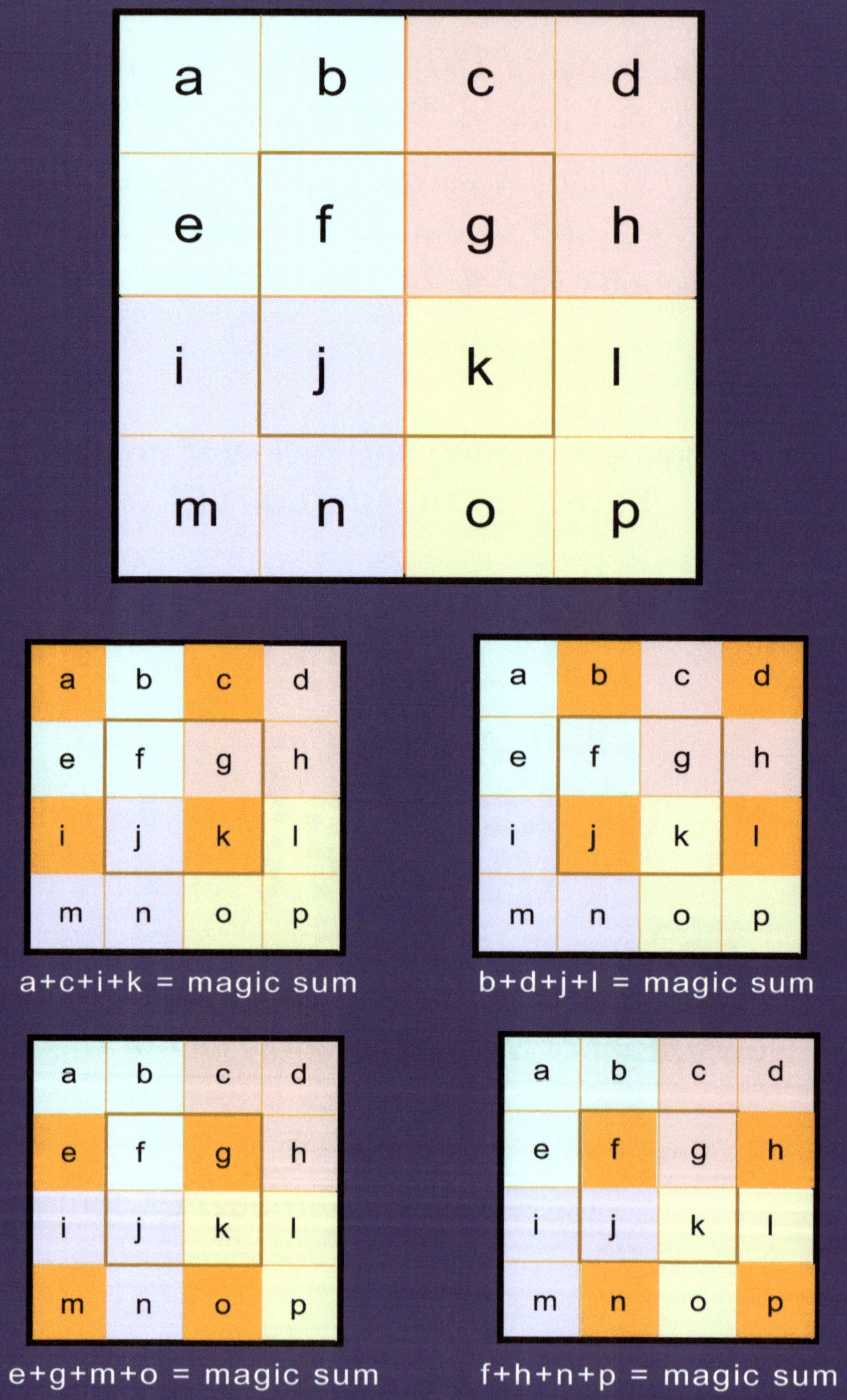

a+c+i+k = magic sum

b+d+j+l = magic sum

e+g+m+o = magic sum

f+h+n+p = magic sum

FIGURE 19

a+c+i+k = b+d+j+l = e+g+m+o = f+h+n+p = magic sum

1	32	36	61
62	35	31	2
63	34	30	3
4	29	33	64

1	32	36	61
62	35	31	2
63	34	30	3
4	29	33	64

1+36+63+30 = 130

1	32	36	61
62	35	31	2
63	34	30	3
4	29	33	64

32+61+34+3 = 130

1	32	36	61
62	35	31	2
63	34	30	3
4	29	33	64

62+31+4+33 = 130

1	32	36	61
62	35	31	2
63	34	30	3
4	29	33	64

35+2+29+64 = 130

FIGURE 20

a	b	c	d
e	f	g	h
i	j	k	l
m	n	o	p

1	32	36	61
62	35	31	2
63	34	30	3
4	29	33	64

e+i+h+l = magic sum   62+63+2+3 = 130

a	b	c	d
e	f	g	h
i	j	k	l
m	n	o	p

1	32	36	61
62	35	31	2
63	34	30	3
4	29	33	64

b+c+n+o = magic sum   32+36+29+33 = 130

FIGURE 21

a	b	c	d
e	f	g	h
i	j	k	l
m	n	o	p

1	32	36	61
62	35	31	2
63	34	30	3
4	29	33	64

a+d+m+p = magic sum   1+61+4+64 = 130

FIGURE 22

The great Indian Mathmatician Srinivasa Ramanujan, discovered some patterns in the 4x4 magic squares, which are, later used by others, to create Birthday Magic Squares. The patterns are given below.

a	b	c	d
e	f	g	h
i	j	k	l
m	n	o	p

$$a + d = n + o$$

a	b	c	d
e	f	g	h
i	j	k	l
m	n	o	p

$$b + c = m + p$$

a	b	c	d
e	f	g	h
i	j	k	l
m	n	o	p

$$a + m = h + l$$

a	b	c	d
e	f	g	h
i	j	k	l
m	n	o	p

$$e + i = d + p$$

a	b	c	d
e	f	g	h
i	j	k	l
m	n	o	p

$$m + d = f + k$$

a	b	c	d
e	f	g	h
i	j	k	l
m	n	o	p

$$a + p = j + g$$

FIGURE 23

1	32	36	61
62	35	31	2
63	34	30	3
4	29	33	64

1+61 = 29+33 = 62

1	32	36	61
62	35	31	2
63	34	30	3
4	29	33	64

32+36 = 4+64 = 68

1	32	36	61
62	35	31	2
63	34	30	3
4	29	33	64

1+4 = 2+3 = 5

1	32	36	61
62	35	31	2
63	34	30	3
4	29	33	64

62+63 = 61+64 = 125

1	32	36	61
62	35	31	2
63	34	30	3
4	29	33	64

4+61 = 35+30 = 65

1	32	36	61
62	35	31	2
63	34	30	3
4	29	33	64

1+64 = 34+31 = 65

FIGURE 24

PUZZLE 1

66
71
119
62
9
109
25
9
120
110
71
71
141
100
8
63
30
89
80
68
61
72
96
46
82
83
26

Start Time :

End Time :

PUZZLE 2

124	69		68	**13**	115
		11	106	25	
62	63		104		106
69		126	152	87	
				70	
	85	41	95		**25**
87		41		103	
		41	82	64	

Start Time :   End Time :

PUZZLE 3

7 | | 126 | 73 | | 120
178 | | | 5 | 113 | 66 | | 12
| | 114 | 19 | | | 67
| 8 | | 187 | 117 | 11
62 | 20 | | | | 29
| 69 | 103 | 56 | 89
| 24 | | 96 | 31
108 | | 58 | | 69

Start Time :

End Time :

PUZZLE 4

	130		67			117
4				**15**		
121		9		116	62	
	60				103	99
69	69	102	146	46		
			70	64		66
87	66				97	**30**
		41	88			**29**

Start Time :

End Time :

PUZZLE 5

	69		67			15	117	
3								
183		9		155		25		
			1					
	87		63					**14**
23		101	149		38	115	92	
	85	63	56		69		96	
47								
87					**32**			
			84		64			**30**
		41						

Start Time :

End Time :

PUZZLE 6

66		62	73		120		
73		61	107			66	66
73			73		59		
	85		138			74	13
19		59					97
89	89		58		73		
		80					99
89			70			27	

Start Time :

End Time :

PUZZLE 7

7

135

114

54

123

69

69

59

90

68

70

11

178

120

3

66

32

69

76

100

48

106

123

26

84

69

22

Start Time :

End Time :

PUZZLE 8

			74		120		
68		58					
122		63		103	27		14
61	61				114		
10	87	143		37			
108					77		
45	85	96			93		26
46		47				96	
82		81		36			

Start Time :

End Time :

PUZZLE 9

62	4	134			52	80	
	66	58		64			
8	121	112		79			10
78	55	67		87		85	
	86				92		
45		81					34
84	87	50		73		67	
		22	111			26	

Start Time :

End Time :

PUZZLE 10

68	2	133			51	69
	73	54		76		50
4		84	131			81
119			131	73		
108					73	
	88	101			97	30
46		53			64	
61		21	105		34	

Start Time :

End Time :

PUZZLE 11

68

89

66

61

53

69

5

101

36

171

6

75

74

90

89

93

137

41

98

57

25

60

69

101

118

24

30

Start Time :

End Time :

PUZZLE 12

63	6	120		10	68	77

183	10	115			
22	112	142	70	64
5	22	112	142	44	87
64					35
67	72	97	94
108				67
63	44	74	27

Start Time :

End Time :

PUZZLE 13

60	6	133			11	118	
	66		60	54	80		
62	104	92					97
72			146	77	116		
	88		88				34
107			18		73	67	
			112	21			26

Start Time :   End Time :

PUZZLE 14

63	5	122			79		62
	121	62			55		
61		61		63			16
23		131	145	88		90	
47	64		96		61		26
66	85	79	43			66	
		48	69		30		

Start Time :

End Time :

PUZZLE 15

63 63 76 109

63

121 64 25

9

59 89 106

62

72 49 131 122 88

61 119 61

26

111 80 66

22

64 66

33

Start Time :

End Time :

PUZZLE 16

		127		80
69	57		52	
121 17		64		
62	55	79		10
25 114	141	115	95	
81				
48	53	111 57	29	
106 66	98		99	
22		27		

Start Time :   End Time :

PUZZLE 17

62

128

4

11

121

71

106

76

170

8

10

102

105

93

87

133

107

93

98

43

34

86

40

92

67

42

30

Start Time :

End Time :

PUZZLE 18

72		60	72		63		56
	117	61		54	25	25	
62		62		99			
22	28	121	147		87	86	
91	87	91		69		27	
		86			95		
21	62	75		33			

Start Time : End Time :

PUZZLE 19

73

72

114

61

54

9

116

69

70

15

106

59

73 21 105 186 45 43

69 104 53 63

32

106

74

99

52

22

27

Start Time :

End Time :

PUZZLE 20

1	68		73		15	119
123			5	159	27	
87	67		55			9
		85	144		50	118
					92	
47	45	103	58			
107				30	93	64
		21	81			40

Start Time :

End Time :

PUZZLE 21

59	8	120		12	63	71
	129		4	104		
91	62		116		67	9
			67		67	114
	89				91	
47	41		82		103	
63	64			29		66
	41		48	69		30

Start Time :

End Time :

PUZZLE 22

63	6	129		52	69
	129		56	27	49
62		24		116	104
71		103	188	42	
	81		93		27
22	67		79	96	66
85		24	78	33	

Start Time :

End Time :

PUZZLE 23

10		71	57	23		52	68
182		9		51	81		
	61			114			
22	81	154			116	90	
		83					
	112		73	63		89	
68			42		63		72
	45	41		30			

Start Time :

End Time :

PUZZLE 24

59		133				107	
	8				**10**		
	70		71		81		**9**
7		122				95	99
127	125	51	104		91		
		106	92	64		105	**34**
							67
23							
61			111				
		22				**26**	

Start Time :

End Time :

PUZZLE 25

66			73		67		
		62					**53**
	69	59			**55**	37	
8		168			63		
81	75	55	92	44	82	91	
90	66		89			57	**32**
		41	120				99
42	64		53		**27**		

<table>
<tr><td>Start Time :</td><td>End Time :</td></tr>
</table>

PUZZLE 26

1	66		70		76		**55**
184		9		107			25
		112	105			118	
4			129		74		
85							**29**
68		100				93	
	85	61		66		63	
44			80				**40**

Start Time :

End Time :

PUZZLE 27

112		81		3	76	113	
	66	64	66	100	49		4
1							115
120	86		161	116			
	97	38		97	97	33	7
38		72					121
49			37	35			

Start Time :

End Time :

PUZZLE 28

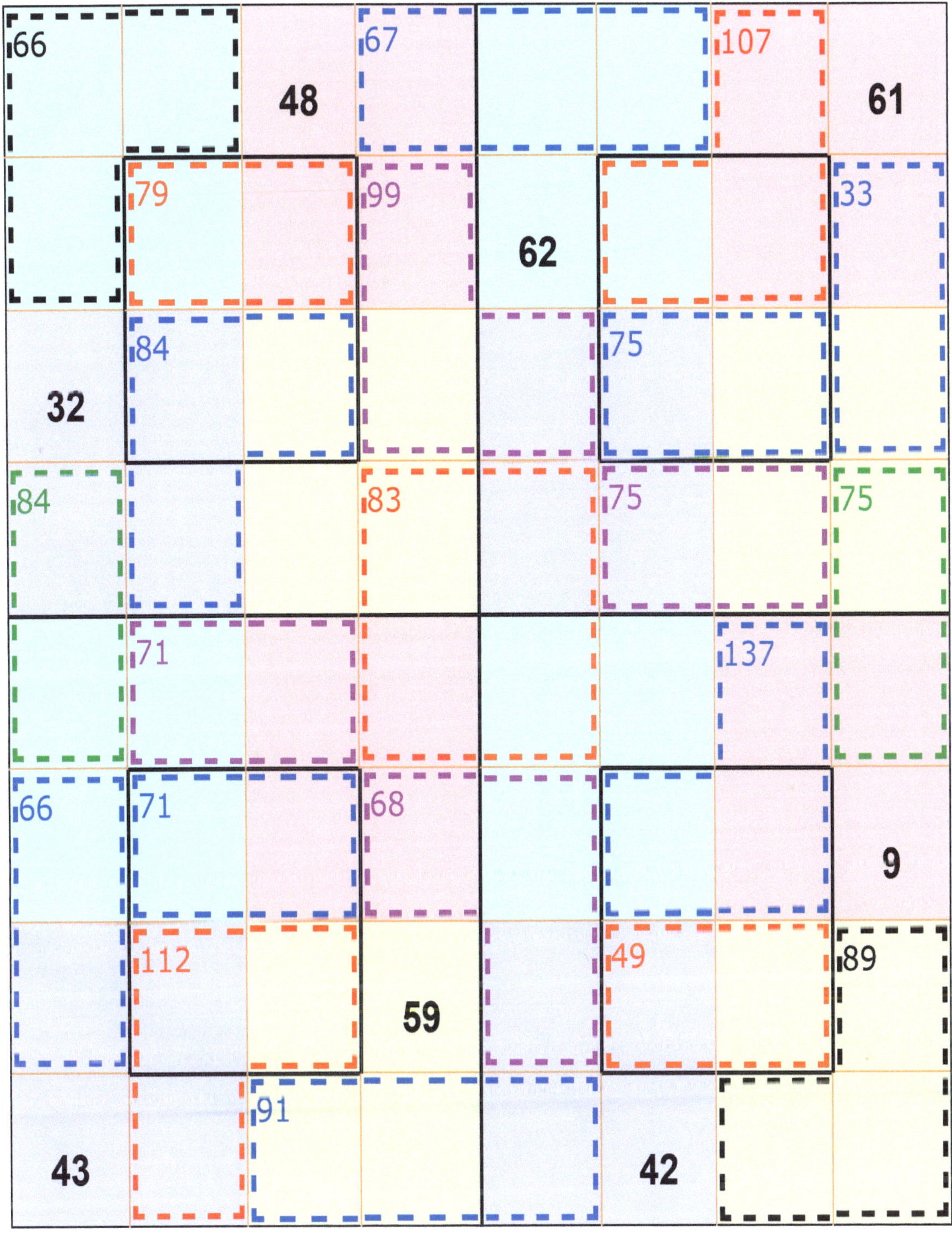

Start Time :

End Time :

PUZZLE 29

47 99 106 81 35

33 97 84 66

82 22 85

7 121 167 61 45

11 64

97 125 125 49

84 104 41

49 44 49

Start Time :

End Time :

PUZZLE 30

63 | 81 | | 67 | 50 | | 100
29
98 | | | | 126
17 | | 87 | | | 62
62 | 107 | 174 | 107 | 60
124 | 73 | | 106
60 | | | | 7
86 | | 60 | 75 | 83
63
43

Start Time :

End Time :

PUZZLE 31

49	32	130	47		30	116
			32	142	85	
50	63					4
64	88		151	70	147	
43	102	55		66	67	7
93					118	83
			100		38	8

Start Time :

End Time :

PUZZLE 32

66
54
100
48
29
81
62
82
100
114
97
49
20
101
108
33
66
107
58
33
88
59
77
80
16
91
101
37
55

Start Time :

End Time :

PUZZLE 33

17   113   78   126

3

50   78   55

83   116   99   74   20

18

98   71   84

66   99

136

59   24

87   96   61   59   26

85

95

11

Start Time :

End Time :

PUZZLE 34

32			70			111	
		48			3		
66	128	79		64			
	35	60		108	50		
						51	
75	78	117				69	
	58				81		
60		53		106		32	
86	71	78					
		63			97		
	43		8				

Start Time :

End Time :

PUZZLE 35

97			35	13	56	116
47	81		161			
88	63	34			64	30
			98		64	114
	71				62	
54	71	121		56		
108	43		64	67		23
		28		106		

Start Time :

End Time :

PUZZLE 36

17		130	49	94			50
63			36	85			
63	97		94	94			19
		131				124	
78			72				
37	118	84					
82		13	122			26	
108		56		49			

Start Time :

End Time :

PUZZLE 37

82 110 3 66 116

47 83 32

86 65 64 80 4

118 48 110

98 138

59 34 79 33

87 64 86 40 26

105

60

<table>
<tr><td>Start Time :</td><td>End Time :</td></tr>
</table>

PUZZLE 38

81		63	82			86	29
	81		63				39
17	117	79	94		96		
55			102		93	101	103
87	34		79				7
	117		61		49		121
38		96			10		

Start Time :   End Time :

PUZZLE 39

96		50		3	104	98	
	83	33					19
48	63			90		99	
82	92		159		103		
		88					
			134		34		9
28	117			97		120	
49		60		8			

Start Time :

End Time :

PUZZLE 40

48	32	83		13	66		94
	97		103				
64		81			64		89
78	12	101	153		28	117	
	64		107			67	9
92		104					82
	66		21	99		7	

Start Time :

End Time :

PUZZLE 41

49

32

110

87

30

114

50

66

29

50

33

67

89

91

46

154

142

117

96

75

8

66

60

83

27

81

119

28

7

Start Time :

End Time :

PUZZLE 42

PUZZLE 43

82

95

34

35

117

115

97

81

16

84

97

30

106

34

106

108

17

89

59

34

152

9

87

59

68

105

11

23

Start Time :

End Time :

PUZZLE 44

112		**50**	64	48		87	
		65			81		**14**
32	85			111		119	
84		77	137		28	128	
	86			103			**9**
53	49		82	80			89
39		**37**			**42**		

Start Time :

End Time :

PUZZLE 1 SOLUTION

50	9	16	55	30	33	27	40
56	15	10	49	39	28	34	29
13	54	51	12	36	31	37	26
11	52	53	14	25	38	32	35
60	3	6	61	41	18	23	48
62	5	4	59	47	24	17	42
7	64	57	2	22	45	44	19
1	58	63	8	20	43	46	21

PUZZLE 2 SOLUTION

50	9	16	55	35	25	38	32
56	15	10	49	40	30	33	27
13	54	51	12	26	36	31	37
11	52	53	14	29	39	28	34
57	2	7	64	45	19	18	48
63	8	1	58	44	22	23	41
6	61	60	3	24	42	43	21
4	59	62	5	17	47	46	20

PUZZLE 3 SOLUTION

51	12	54	13	29	34	28	39
53	14	52	11	40	27	33	30
16	55	9	50	35	32	38	25
10	49	15	56	26	37	31	36
63	5	4	58	47	19	42	22
60	2	7	61	46	18	43	23
6	64	57	3	20	48	21	41
1	59	62	8	17	45	24	44

PUZZLE 4 SOLUTION

54	12	9	55	35	28	38	29
51	13	16	50	37	30	36	27
15	49	52	14	32	39	25	34
10	56	53	11	26	33	31	40
57	2	7	64	45	19	18	48
63	8	1	58	44	22	23	41
6	61	60	3	24	42	43	21
4	59	62	5	17	47	46	20

PUZZLE 5 SOLUTION

52	11	14	53	39	27	34	30
54	13	12	51	38	26	35	31
15	56	49	10	28	40	29	33
9	50	55	16	25	37	32	36
58	1	8	63	45	19	18	48
64	7	2	57	44	22	23	41
5	62	59	4	24	42	43	21
3	60	61	6	17	47	46	20

PUZZLE 6 SOLUTION

51	12	54	13	39	32	25	34
53	14	52	11	33	26	31	40
16	55	9	50	30	37	36	27
10	49	15	56	28	35	38	29
63	3	58	6	48	23	18	41
62	2	59	7	42	17	24	47
4	64	5	57	21	46	43	20
1	61	8	60	19	44	45	22

PUZZLE 7 SOLUTION

13	50	11	56	37	27	26	40
51	16	53	10	32	34	35	29
54	9	52	15	36	30	31	33
12	55	14	49	25	39	38	28
63	5	4	58	46	18	43	23
60	2	7	61	47	19	42	22
6	64	57	3	17	45	24	44
1	59	62	8	20	48	21	41

PUZZLE 8 SOLUTION

52	14	9	55	38	26	35	31
53	11	16	50	39	27	34	30
15	49	54	12	25	37	32	36
10	56	51	13	28	40	29	33
64	7	2	57	45	19	18	48
58	1	8	63	24	42	43	21
5	62	59	4	44	22	23	41
3	60	61	6	17	47	46	20

PUZZLE 9 SOLUTION

53	11	10	56	29	34	27	40
16	50	51	13	35	32	37	26
52	14	15	49	38	25	36	31
9	55	54	12	28	39	30	33
63	3	58	6	21	42	20	47
62	2	59	7	48	19	41	22
4	64	5	57	43	24	46	17
1	61	8	60	18	45	23	44

PUZZLE 10 SOLUTION

13	50	12	55	36	30	25	39
56	11	49	14	37	27	32	34
51	16	54	9	31	33	38	28
10	53	15	52	26	40	35	29
59	1	62	8	45	19	18	48
64	6	57	3	24	42	43	21
2	60	7	61	44	22	23	41
5	63	4	58	17	47	46	20

PUZZLE 11 SOLUTION

53	11	10	56	37	25	32	36
52	14	15	49	40	28	29	33
16	50	51	13	27	39	34	30
9	55	54	12	26	38	35	31
64	4	57	5	47	21	20	42
61	1	60	8	44	18	23	45
2	62	7	59	22	48	41	19
3	63	6	58	17	43	46	24

PUZZLE 12 SOLUTION

52	14	9	55	35	28	38	29
53	11	16	50	37	30	36	27
15	49	54	2	32	39	25	34
10	56	51	13	26	33	31	40
63	8	1	58	45	17	24	44
57	2	7	64	48	20	21	41
6	61	60	3	19	47	42	22
4	59	62	5	18	46	43	23

PUZZLE 13 SOLUTION

53	9	16	52	29	34	27	40
56	12	13	49	35	32	37	26
11	55	50	14	38	25	36	31
10	54	51	15	28	39	30	33
63	5	4	58	45	19	18	48
60	2	7	61	24	42	43	21
6	64	57	3	44	22	23	41
1	59	62	8	17	47	46	20

PUZZLE 14 SOLUTION

53	9	16	52	38	26	35	31
56	12	13	49	39	27	34	30
11	55	50	14	25	37	32	36
10	54	51	15	28	40	29	33
64	7	2	57	45	19	18	48
58	1	8	63	24	42	43	21
5	62	59	4	44	22	23	41
3	60	61	6	17	47	46	20

PUZZLE 15 SOLUTION

PUZZLE 16 SOLUTION

PUZZLE 17 SOLUTION

56	15	10	49	29	34	28	39
50	9	16	55	40	27	33	30
13	54	51	12	35	32	38	25
11	52	53	14	26	37	31	36
63	3	58	6	47	24	17	42
62	2	59	7	41	18	23	48
4	64	5	57	22	45	44	19
1	61	8	60	20	43	46	21

PUZZLE 18 SOLUTION

56	15	10	49	37	27	26	40
50	9	16	55	32	34	35	29
13	54	51	12	36	30	31	33
11	52	53	14	25	39	38	28
61	3	2	64	44	22	17	47
60	6	7	57	45	19	24	42
8	58	59	5	23	41	46	20
1	63	62	4	18	48	43	21

PUZZLE 19 SOLUTION

13	50	11	56	39	32	25	34
51	16	53	10	33	26	31	40
54	9	52	15	30	37	36	27
12	55	14	49	28	35	38	29
60	6	1	63	46	18	43	23
61	3	8	58	47	19	42	22
7	57	62	4	17	45	24	44
2	64	59	5	20	48	21	41

PUZZLE 20 SOLUTION

52	14	9	55	29	34	27	40
53	11	16	50	35	32	37	26
15	49	54	12	38	25	36	31
10	56	51	13	23	39	30	33
63	5	4	58	45	19	18	48
60	2	7	61	24	42	43	21
6	64	57	3	44	22	23	41
1	59	62	8	17	47	46	20

PUZZLE 21 SOLUTION

55	16	9	50	38	26	35	31
49	10	15	56	39	27	34	30
14	53	52	11	25	37	32	36
12	51	54	13	28	40	29	33
59	4	62	5	21	42	19	48
61	6	60	3	43	24	45	18
8	63	1	58	46	17	44	23
2	57	7	64	20	47	22	41

PUZZLE 22 SOLUTION

14	49	11	56	37	27	26	40
55	12	50	13	36	30	31	33
52	15	53	10	32	34	35	29
9	54	16	51	25	39	38	28
57	2	7	64	48	20	41	21
63	8	1	58	45	17	44	24
6	61	60	3	18	46	23	43
4	59	62	5	19	47	22	42

PUZZLE 23 SOLUTION

55	13	12	50	40	31	26	33
52	10	15	53	34	25	32	39
14	56	49	11	29	38	35	28
9	51	54	16	27	36	37	30
57	2	7	64	47	19	42	22
63	8	1	58	46	18	43	23
6	61	60	3	20	48	21	41
4	59	62	5	17	45	24	44

PUZZLE 24 SOLUTION

51	9	54	16	29	34	27	40
56	14	49	11	35	32	37	26
10	52	15	53	38	25	36	31
13	55	12	50	28	39	30	33
59	4	62	5	21	42	20	47
61	6	60	3	48	19	41	22
8	63	1	58	43	24	46	17
2	57	7	64	18	45	23	44

PUZZLE 25 SOLUTION

53	9	16	52	39	32	25	34
56	12	13	49	33	26	31	40
11	55	50	14	30	37	36	27
10	54	51	15	28	35	38	29
63	3	58	6	43	17	46	24
62	2	59	7	48	22	41	19
4	64	5	57	18	44	23	45
1	61	8	60	21	47	20	42

PUZZLE 26 SOLUTION

55	13	12	50	29	34	27	40
52	10	15	53	35	32	37	26
14	56	49	11	38	25	36	31
9	51	54	16	28	39	30	33
61	3	2	64	21	42	20	47
60	6	7	57	48	19	41	22
8	58	59	5	43	24	46	17
1	63	62	4	18	45	23	44

PUZZLE 27 SOLUTION

52	4	45	29	41	7	26	56
61	13	36	20	58	24	9	39
14	62	19	35	23	57	40	10
3	51	30	46	8	42	55	25
17	34	32	47	60	12	21	37
64	15	49	2	53	5	28	44
33	18	48	31	6	54	43	27
16	63	1	50	11	59	38	22

PUZZLE 28 SOLUTION

61	13	20	36	39	9	58	24
52	4	29	45	56	26	41	7
3	51	46	30	25	55	8	42
14	62	35	19	10	40	23	57
50	17	47	16	38	5	59	28
48	15	49	18	60	27	37	6
31	64	2	33	11	44	22	53
1	34	32	63	21	54	12	43

PUZZLE 29 SOLUTION

19	36	30	45	56	8	41	25
62	13	51	4	57	9	40	24
35	20	46	29	10	58	23	39
14	61	3	52	7	55	26	42
49	31	2	48	60	12	21	37
34	16	17	63	53	5	28	44
32	50	47	1	6	54	43	27
15	33	64	18	11	59	38	22

PUZZLE 30 SOLUTION

35	13	62	20	40	7	57	26
52	30	45	3	58	25	39	8
29	51	4	46	9	42	24	55
14	36	19	61	23	56	10	41
33	2	64	31	59	11	22	38
63	32	34	1	54	6	27	43
18	49	15	48	5	53	44	28
16	47	17	50	12	60	37	21

PUZZLE 31 SOLUTION

51	29	4	46	40	7	57	26
36	14	19	61	58	25	39	8
30	52	45	3	9	42	24	55
13	35	62	20	23	56	10	41
34	15	17	64	44	5	22	59
63	18	16	33	53	28	11	38
32	49	47	2	27	54	37	12
1	48	50	31	6	43	60	21

PUZZLE 32 SOLUTION

46	3	20	61	40	7	57	26
51	30	13	36	58	25	39	8
29	52	35	14	9	42	24	5
4	45	62	19	23	56	10	41
50	17	47	16	43	21	6	60
48	15	49	18	54	12	27	37
31	64	2	33	28	38	53	11
1	34	32	63	5	59	44	22

PUZZLE 33 SOLUTION

61	13	20	36	41	7	26	56
52	4	29	45	58	24	9	39
3	51	46	30	23	57	40	10
14	62	35	19	8	42	55	25
64	16	33	17	27	37	6	60
49	1	48	32	44	22	53	11
15	63	18	34	54	12	43	21
2	50	31	47	5	59	28	38

PUZZLE 34 SOLUTION

46	4	51	29	40	9	23	58
61	19	36	14	57	24	10	39
3	45	30	52	26	55	41	8
20	62	13	35	7	42	56	25
50	17	47	16	59	11	22	38
48	15	49	18	54	6	27	43
31	64	2	33	5	53	44	28
1	34	32	63	12	60	37	21

PUZZLE 35 SOLUTION

51	3	30	46	58	10	23	39
62	14	19	35	55	7	26	42
4	52	45	29	8	56	41	25
13	61	36	20	9	57	40	24
33	2	64	31	38	5	59	28
63	32	34	1	60	27	37	6
18	49	15	48	11	44	22	53
16	47	17	50	21	54	12	43

PUZZLE 36 SOLUTION

36	14	19	61	23	40	26	41
51	29	4	46	58	9	55	8
30	52	45	3	39	24	42	25
13	35	62	20	10	57	7	56
49	1	32	48	53	27	6	44
64	16	17	33	38	12	21	59
2	50	47	31	28	54	43	5
15	63	34	18	11	37	60	22

PUZZLE 37 SOLUTION

45	19	4	62 (110)	41	7	26	56
52 (116)	14	29	35	58	24 (33)	9	39
30 (66)	36	51 (80)	13 (48)	23 (138)	57	40	10 (105)
3	61	46	20	8	42	55	25
48 (110)	1	18 (64)	63 (118)	27	37 (79)	6 (86)	60
49	32	15 (65)	34	44	22	53 (64)	11
31	50 (83)	33	16	54 (98)	12 (34)	43	21
2 (82)	47	64 (86)	17	5	59	28 (87)	38

PUZZLE 38 SOLUTION

29	35 (39)	4	62 (103)	41	7	26 (121)	56
52 (86)	14	45	19 (101)	58	24	9	39
46	20	51 (96)	13 (93)	23	57	40 (49)	10
3	61	30	36	8	42	55	25
33 (82)	2 (63)	64 (94)	31 (102)	27	37 (79)	6 (61)	60
63	32	34	1	44	22	53	11 (96)
18	49 (81)	15 (117, 79)	48	54	12 (34)	43 (117)	21
16 (81)	47	17	50 (55)	5	59 (87)	28	38

PUZZLE 39 SOLUTION

PUZZLE 40 SOLUTION

PUZZLE 41 SOLUTION

51	29	4	46	39	8	58	25
36	14	19	61	57	26	40	7
30	52	45	3	24	55	9	42
13	35	62	20	10	41	23	56
34	15	17	64	60	12	37	21
63	18	16	33	53	5	44	28
32	49	47	2	11	59	22	38
1	48	50	31	6	54	27	43

PUZZLE 42 SOLUTION

46	4	51	29	58	10	23	39
61	19	36	14	55	7	26	42
3	45	30	52	8	56	41	25
20	62	13	35	9	57	40	24
49	1	32	48	37	22	12	59
64	16	17	33	60	11	21	38
2	50	47	31	27	44	54	5
15	63	34	18	6	53	43	28

PUZZLE 43 SOLUTION

19	36	30	45	57	9	24	40
62	13	51	4	56	8	25	41
35	20	46	29	7	55	42	26
14	61	3	52	10	58	39	23
34	1	63	32	27	37	6	60
64	31	33	2	44	22	53	11
15	48	18	49	54	12	43	21
17	50	16	47	5	59	28	38

PUZZLE 44 SOLUTION

36	14	19	61	39	9	58	24
51	29	4	46	56	26	41	7
30	52	45	3	25	55	8	42
13	35	62	20	10	40	23	57
31	33	2	64	43	5	28	54
50	16	47	17	60	22	11	37
48	18	49	15	21	59	38	12
1	63	32	34	6	44	53	27

Register For Free Training For Magic Puzzles

https://forms.gle/sRR1hww9y2gQsvrt9

www.ingramcontent.com/pod-product-compliance
Lightning Source LLC
Chambersburg PA
CBHW040044240726
48664CB00004B/1063